Space Tourism and Jesus

Unity of Mankind Beyond Earth

Table of Contents

Chapter 1. Introduction

Explore the fascinating intersection of technology, theology, and human ambition in our special report: "Space Tourism and Jesus: Unity of Mankind Beyond Earth". As humankind takes its first tentative steps towards becoming a space-faring species, this report delves into how our religious beliefs, particularly focusing on the teachings of Jesus Christ, coincide with this unprecedented venture. Whether it's the unity of mankind beyond the planet or the infinite love that spans galaxies, this report promises to transport you to a realm where spirituality meets the stars. Brimming with thought-provoking discourse, delightful insights, and heartwarming anecdotes, our report is a testament to humanity's relentless pursuit of understanding and harmony. It's an adventurous leap into the unexplored that will leave you intrigued, informed, and longing for more. A cosmic journey awaits you; engage with this rewarding narrative and let it inspire the stargazer and philosopher within.

Chapter 2. Transition into the Cosmos

Over the millennia, humans have gazed up at the star-dotted sky with reverence, curiosity, and longing. The eternal vastness of the cosmos seemed unreachable, a realm exclusively reserved for divine beings. Yet in the last century, advances in technology have made it possible for mankind to cross the threshold of our home—the Earth. We've set foot on the moon, sent probes to the furthest corners of our solar system, and now are on the precipice of venturing further, courtesy of space tourism.

2.1. Beyond The Blue Sky

The transition into the cosmos commenced in earnest with the inception of the space age in the mid-20th century. As Yuri Gagarin, the first human in space, rocketed into the cosmos on April 12, 1961, he brought humanity along with him, forever shifting our perspective from "Earthbound" to "cosmic".

The famous photograph taken by the Apollo 8 crew, dubbed "Earthrise", showed a fragile blue marble suspended in the vast darkness of space, further reinforcing this shift. The sight drove home the profound understanding of the unity, fragility, and preciousness of life on Earth in a boundless universe.

Space travel propelled our species from observing distant celestial bodies through a lens to examining them hands-on. The moon was no longer only a romantic feature of a midsummer night's sky, but a place where human footprints were etched into its silent regolith.

2.2. Unity And Space Exploration

The instinct to explore is buried deep within our DNA. It was by crossing hazardous oceans, huge deserts, and frigid icecaps, that homo sapiens made Earth their home. This unified effort from early human ancestors showcases the innate human capacity for cooperative endeavors—an essential facet of Jesus' teachings.

The same unity is now allowing humanity to venture into the cosmos. International institutions like the European Space Agency and partnerships between nations showcase the pooling of resources and collective vision. The International Space Station—a result of global collaboration—is a testament to this unity, symbolizing peaceful coexistence and shared goals.

2.3. Space Tourism Envisioned

In this modern age, the idea of space tourism—a trip to space purely for enjoyment or non-scientific purposes—is gradually transitioning from a dream into reality. Several enterprising companies, including SpaceX, Blue Origin, and Virgin Galactic, are making concerted efforts to make space travel accessible not just for astronauts but for ordinary individuals.

Space tourism, the concept of any person being able to look down on Earth from space, is undoubtedly an exemplification of humanity's progress, a pursuit that Jesus' teachings of humility, resilience, and ambition can resonate with.

2.4. Pondering on Divinity in the Vast Cosmos

As we venture more into space, the infinite expanse could offer fresh perspectives on theology too. God's creation, as per Christian teachings, is viewed as awe-inspiring and perfect. Seeing the universe firsthand could lead to a renewed affirmation in believers for the Divine hand in the wondrous creation.

However, the infinite mysteries of the cosmos could also stir questions. Perhaps challenges to established beliefs could establish innovative spiritual insights. Will we redefine divinity as we delve deeper into space? Will our understanding of God and Jesus' teachings evolve as we uncover the secrets of the universe? These curiosities deliver a fresh dimension to faith and can reinvigorate man's ongoing spiritual journey.

2.5. The Future of Space Travel and Spirituality

The future of deep-space exploration promises to introduce a new paradigm of unity through space tourism. The Church, an institution rooted in communal unity and agape, will have a valuable role to play in this transition, shaping the narrative of space exploration with its principles while adapting to ambiguities and mysteries that await us in space.

By looking at the teachings of Jesus—focused on love, humility, unity, and the pursuit of knowledge—we find they harmonize with the vision for space travel and the collective human ambition. The transition into the cosmos is by no means a deviation from these teachings but rather a realization of them on a grand, cosmic scale.

Could it be that space tourism may stimulate a spiritual revolution, where the clarity of our place in the universe leads to a newfound inter-connectedness? An enhanced perspective of brotherly love and unity for the entirety of humanity? Will each subsequent journey into the cosmos spark more questions that lead to self-discovery and enlightenment? The interplay of these notions beckons towards an exciting future in humankind's transcendence to becoming a space-faring species.

The venture into space bridges the gap between human ambition and divine design. As mankind reaches out to the stars, embracing our smallness in the vast cosmos, we find ourselves in a deeply spiritual dialogue with the universe, an echo of Jesus' teachings of love and unity captured by technology's mirror held up against the infinite.

Chapter 3. Space Tourism: The Final Frontier

Since time immemorial, mankind has gazed at the night sky with fascination, marveling at the vastness of the universe and dreaming of traversing to distant celestial bodies. Now, as we stand at the precipice of a new era, these distant dreams are gradually evolving into reality. The final frontier, once truly inaccessible, is slowly opening its gates to humankind, not just astronauts and scientists, but to common individuals, transforming passive observers into active participants in the cosmic odyssey. This is the dawn of space tourism, fuelling our innate penchant for exploration and aspiration to transcend the bonds of Earth. It represents the apotheosis of technological advancement, particularly in aerospace engineering.

3.1. The Advent of Space Tourism

The term 'space tourism' comes across as an enticing oxymoron, blending celestial escapade with the leisurely travel fondly associated with vacationing on Terra. To understand this surreal concept, one must trace back to the mid-20th century when Russia and America were locked in a tense space race. Yuri Gagarin's voyage in 1961 as the first human in space introduced the possibility of space travel. However, those pioneers were not tourists, but steely professionals zooming through the cosmos for their nations' strategic interests.

Advancements in space technologies from the 1950s onwards sparked a fascination with space, leading to a gradually increasing, yet largely unfulfilled, desire for non-astronauts to experience such an odyssey. The first major space tourist, American businessman Dennis Tito, did not break away from Earth's gravity until 2001, traveling to the International Space Station aboard a Russian Soyuz spacecraft.

Undoubtedly, Tito's journey was a striking landmark in the evolution of space travel. It was when "outer space" was reframed from an exclusive playground for nations' strategic interests into a potential tourism hotspot. Herein lie the challenges and opportunities for humanity at the precipice of this Final Frontier.

3.2. The Technological Triumph

While the idea of space tourism is steeped in human ambition and curiosity, its feasibility has been ushered in by relentless technological innovation. Reusable rockets, technological efficiency, and evolving safety measures, spearheaded by companies like SpaceX, Blue Origin, and Virgin Galactic, are easing the accessibility of outer space, gradually reducing costs, and increasing reliability.

SpaceX's innovative reusable rocket technology has been particularly transformative in reducing costs. Unlike traditional rockets, which are single-use and thus exceedingly expensive, SpaceX's Falcon and Starship series can return to Earth and be repurposed for multiple missions.

Blue Origin, too, is opening new avenues with its sub-orbital space tourism technology. The New Shepard is a reusable vehicle expressly designed for space tourism, engineered to transport up to six passengers just beyond the edge of space, allowing them to experience a few minutes of weightlessness.

Virgin Galactic, meanwhile, seeks to pioneer a different path with its unique air-launch system that uses a custom-built carrier aircraft to transport a spacecraft to high altitudes, where it then separates and ignites its rocket engines. This method eliminates the need for a traditional launch pad and allows for a smoother, safer journey to space.

3.3. The Psychological Thrust

Conversely, the psychological aspects of space tourism have also come into play as the Final Frontier has become more accessible. The ability to witness Earth from space, referred to as the "overview effect," is said to provoke a cognitive shift in perspective, bound to have profound effects.

Behavioral scientists theorize that this expansive glimpse from space elicits a deep sense of unity and interconnectedness. As individuals gaze back at our solitary planet suspended in the void, borders and geopolitical divisions blur, fostering global unity and a heightened sense of empathy among space tourists. This is not just a scientific journey, but also a transformative spiritual experience.

However, further studies are needed to understand the implications of long-duration space travel on human psychology. As private space companies aim for long-haul trips and vacations on Lunar and Martian surfaces, the psychological preparedness of layman astronauts assumes significant importance.

3.4. Forging Ahead

As exhilarating as it may seem, the path forward for space tourism is not without challenges. The endeavor is fraught with technical difficulties, astronomical costs, physiological effects of space travel, and ethical conundrums. Space tourism, like any pioneering field, must navigate these obstacles adeptly.

In addition, the democratization of space comes with the responsibility of ensuring its preservation. With increasing traffic, the problem of space debris is escalating. Stakeholders must devise responsible practices to prevent orbit pollution and truly sustainable space tourism.

Indeed, as technology evolves and makes space gradually more accessible, legal, regulatory, and ethical frameworks also need to be developed and refined. The final frontier, as it welcomes more humans, cannot be a lawless expanse but a domain that is regulated, safe, and equitable.

Humanity stands at a crossroads of a grand cosmic narrative that could redefine our relationship with the cosmos. Space tourism embodies our continuing ambition to explore, discover, and ultimately understand our place in the universe. As we step ahead to traverse the eternal void, we must move conscientiously and cooperatively, ensuring that the final frontier remains a beacon of opportunities, not just for us, but for generations to come.

Chapter 4. Jesus and the Cosmos: Biblical Interpretations

For centuries, we have looked to the skies in wonder, a part of the vast cosmos and constrained only by our understanding. As we take steps toward exploring the universe, an inevitable question arises: How does Christianity, particularly the teachings of Jesus Christ, interpret our place in the cosmos? A thorough examination of the Bible, aided by keen theological insight, facilitates our exploration of this topic.

4.1. God's Vast Creation

The Bible starts with a straightforward representation of cosmic creation in Genesis. "In the beginning, God created the heavens and the earth" (Genesis 1:1). This simple yet profound statement suggests a cosmic perspective of creation. The 'heavens' reflect the entirety of the cosmos beyond our terrestrial sphere. To believers, God is not a local deity but the Creator of the universe in its undisputed expansive glory.

4.2. Christ: The Bond of the Cosmos

Colossians 1:16-17 show Christ's profound connection to the cosmos: "For by him all things were created: things in heaven and on earth, visible and invisible, whether thrones or powers or rulers or authorities; all things have been created through him and for him. He is before all things, and in him all things hold together."

These verses establish Christ not just as a spiritual linchpin, but as an essential, unifying force that permeates our reality and the cosmos at large. They elevate Jesus from being just an earthly figure to cosmic significance, serving as a channel for God's love to flow through the universe.

4.3. The Far Reaching Love of God

Often we view God's love as a terrestrial concept experienced within our microcosm of existence. However, the scriptures speak of a divine love that is as expansive as the universe. Ephesians 3:18-19 states, "May you have the power to understand, as all God's people should, how wide, how long, how high, and how deep his love is. May you experience the love of Christ, though it is too great to understand fully."

This verse signifies God's love is not confined by terrestrial bounds; it spans across the width and length of the Earth, reaching the heights of the skies, and extending as deep as the oceans and beyond. It is a love that is at once all-encompassing and uniquely individual, bringing unity amid diversity, a quality that humanity needs to emulate if it ventures into space.

4.4. Promise of a New Heaven and New Earth

The scriptures also prophesy a future transformation of the cosmos. Revelations 21:1 says, "Then I saw a new heaven and a new earth, for the old heaven and the old earth had disappeared." This apocalyptic vision offers hope of a future reality that encompasses not just Earth, but the entirety of the cosmos.

4.5. The Parable of the Mustard Seed

A revisiting of the parable of the mustard seed paves the way toward understanding the responsibility entrusted to humanity for the cosmos. As Jesus stated in Matthew 13:31-32, the kingdom of heaven is like a mustard seed, which a man took and planted in his field. It is the smallest of all seeds, yet when it grows, it is the largest of garden plants and becomes a tree, so that the birds of the air come and nest in its branches.

This simple narrative holds much philosophical weight. It suggests that the smallest slice of heaven, when implanted on Earth (the field), can grow to such a vast extent that it provides shelter for life (the birds). This directly parallels the notion of mankind's space exploration—the smallest presence of humanity in space (the mustard seed), when carefully nurtured, has potential to grow to provide shelter for life beyond Earth.

4.6. The Cosmic Perspective of the Great Commission

Jesus's Great Commission (Matthew 28:19-20) has been interpreted by many as a mandate for evangelism across the terrestrial globe. However, its implications could be taken one step further to consider a cosmic perspective: "Go, therefore, and make disciples of all nations, baptizing them in the name of the Father and of the Son and of the Holy Spirit."

In conclusion, biblical interpretations suggest a grand interplay between Jesus Christ, humanity, and the cosmos. They signify our expanding roles, not as conquerors of the cosmos, but as stewards and active participants in the cosmic ecosystem. It challenges us to carry not just our ambitions but also our terrestrial lessons of love, unity, and stewardship into the broad expanse of the cosmos.

Chapter 5. Unity Beyond Earth: Christ's Message in the Universe

The inception of the space age raised many questions about our role as a species in the cosmos. It invigorated our quest to seek answers, not just scientific or technical, but also philosophical and spiritual. Many found themselves confronted with an unexpected intersection of faith and space exploration.

5.1. Faith: Transcending the Worldly

Arguably, faith has always been a cosmic endeavor. Religions worldwide have often drawn from astronomical observations to frame the earth and human existence within the larger context of the cosmos. The teachings of Jesus Christ in Christianity also resonate with this aspirational dimension of our existence.

Jesus came with the message of love and unity - principles that were made to transcend political borders, and ultimately imbue humanity with a unifying spirit. The universality of his teachings – to love thy neighbour, to help the needy, and to seek wisdom – can be wonderfully applied to a widening scope, as we extend the 'neighbourhood' to our entire cosmic surroundings, and neighbours to alien life forms, not yet known but conceivable.

The Earth, as viewed from space, offers a boundary-less unity that resonates with one of the pivotal teachings of Jesus Christ - the unity of mankind. Astronaut Edgar Mitchell once famously mentioned encountering an overwhelming sense of universal connectedness when he looked down at the Earth from lunar orbit. He said, "You develop an instant global consciousness – a people orientation, an intense dissatisfaction with the state of the world, and a compulsion to do something about it." His experience aligns strikingly with Christ's emphasis on love and unity among all.

5.2. Ethics in the Cosmos: Guiding Space Travel and Settlement

As humanity ventures into space—with the goals of exploration, colonization, or tourism—our moral compass should not be left behind. In these nascent stages of becoming a space-faring civilization, ethical considerations should guide our actions. Christ's teachings on justice, love, and respect for all creation can shed light on handling unforeseen ethical conundrums.

Our actions in space should be guided by love, not territorial expansion or resource acquisition. We must approach astrobiology and the potential discovery of extraterrestrial life with utmost humility, fostering unity within our kind and promoting peaceful coexistence instead of imposing our belief systems or damaging potentially life-sustaining ecosystems in space.

We must also take into account fairness in space access and benefits—avoiding space becoming a playground solely for wealthy tourists or intense geopolitical competition. The blessings of space – its vastness, mystery, and resources – should belong to all humankind, reflecting Christ's concern for social justice.

5.3. Christianity and Cosmic Pluralism

The discovery of extraterrestrial life, whether microbial or sentient, would trigger profound theological discussions. Some might see this as challenging the uniqueness of human life and God's incarnation in Jesus Christ. However, looking theologically, these views might be limited, constraining God's creation to our planetary understanding.

Assuming sentient extraterrestrial life exists, we can speculate they likely have their own experiences of the divine and their expressions of faith, challenging us to broaden our perspectives, leading to a more nuanced understanding of the divine. Jesus' message of universal love and acceptance would enable us to embrace such revelations without fear, but instead with an open heart and mind.

5.4. The Church and Space

As the intersection of theology and astrophysics becomes increasingly significant, churches have a role to play in the discourse on space exploration. They can encourage congregants to engage with the emerging challenges, to participate and contribute, grounding responses in Christ's teachings of love, unity, and compassion.

All the while, churches should continue to champion the protection of Earth, standing against any space endeavours that harm our planet in favour of short-term, parochial interests.

The overarching theme of this exploration is unity—by observing our planet from space, we can truly grasp the interconnectedness of our world. The teaching of Jesus to "Love thy neighbour" no longer constrains us to geographical proximity but opens a pandora's box of cosmic proportions.

As we progress and the cosmic neighbourhood expands, if we root ourselves in the teachings of Jesus Christ, celebrating the unity and embracing the cosmic pluralism, we would remain truly human in the infinite expanse. Deeper understanding of our universe may pave the way for a deeper understanding of our Maker. After all, the heavens have been declaring the glory of God long before humanity set its reach for them.

Entering the space age doesn't mean leaving behind the profound and fundamental teachings of figures such as Jesus Christ. Instead, they become even more critical as we enter the vast expanse of space. A grain of humility, a sense of unity, and a heart filled with love – these seem to be crucial elements for a meaningful journey in outer space, as on Earth. Let's hold them close as we venture out into the cosmos.

Chapter 6. Science and Religion: The Convergence in the Expanse

In the boundless ocean of cosmic existence, faith and science may seem at odds. Yet, their paths converge when the curtains of human existence are pulled back to reveal the broader stage upon which life plays out; the celestial theater.

6.1. Theological Perspectives on the Cosmos

The religious foundation of Christianity pivots on the teachings of Jesus Christ, centered around love, unity, tolerance, and the eternal quest for knowledge. As we advance technologically and set sights on the cosmos, we encounter exigent questions of theology. Does the universe, boundless and inscrutable, resonate with Christ's teachings? How does Christ's message of boundless love glisten amidst the canvas of infinite galaxies?

The vast expanse of space encapsulates a paradox that resonates with Christ's teachings. Consider a celestial body thousands of light-years away. Its light takes millennia to reach us, conveying a vision of the star as it was, not as it is. This is a stark reflection of Christ's teachings on love and forgiveness; we must love and forgive not merely for what is seen now, but also for what was in the past.

6.2. Quantum Physics and Miracles

Science breaks ground when it comes to rationalizing miracles, a prevalent concept in Christian theology. Quantum physics introduces the observer effect, asserting the act of observance intrinsically alters the observed phenomena. In religious terms, this might imply that divine miracles could be God's observation, impacting reality on a quantum level. This agreement between science and faith paints a picture of harmony, where both strive to grasp the mysteries of existence.

6.3. The Search for Extraterrestrial Life

The ambitious search for extraterrestrial life highlights another convergence. Should we discover alien life, it would be a testament to God's infinite creativity, His boundless love accommodative of lifeforms alien to our understanding. Yet, it also raises theological questions around redemption and salvation. Would Christ's salvation be universal, stretching across all sentient life in the cosmos? This question echoes science's pursuit of life beyond Earth, while encouraging the dialogue with theology on universal redemption.

6.4. Unity in Diversity: A Leap Towards Cosmic Citizenship

For centuries, differences in creed, race, and cultures have carved chasms among humans. But, as we turn our faces towards the stars, we inch closer to embracing a new title: Cosmic Citizens. Observing our planet from space, astronauts uphold the profound unity of mankind - living embodiments of Christ's teaching of universal love and unity, free from terrestrial boundaries. Famously, Apollo 8 astronaut, Bill Anders, reflected, "We came all this way to explore the moon, and the most important thing is that we discovered the Earth."

6.5. The Question of a Cosmic Christ

Christian theologians have broached the 'Cosmic Christ' concept, viewing Christ as an embodiment of the divine present in every corner of the Universe. From the microcosm of the atomic world to the macrocosm of galaxy clusters, His presence permeates existence. This notion marries Christ's message of all-encompassing love with the scientific understanding of the universe's interconnectedness.

6.6. Theological Implications of Deep Time

The understanding of 'Deep Time', the concept that billions of years precede us and billions more will follow, can humble us, reminding us of our temporality and responsibilities. It echoes the teachings of Jesus Christ about humility and stewardship. For Christians, the concept of stewardship extends beyond our planet to the cosmos we may inhabit in the future. Thus, the astronomical projection of the far future demands an ethico-religious response and promotes dialogue between science and theology.

Dichotomous as they seem, science and theology seek to decode the same mystery: the nature of our existence. In this shared pursuit, as mankind steps off the planet and leaps towards the stars, it is critical to foster this harmonious dialogue. The quest does not lie in choosing between faith and reason, but reconciling the two. In the conversation of faith and astrophysics, the teachings of Jesus Christ light the way, offering guidance to navigate the awe-inspiring cosmos while holding fast to our shared humanity.

Chapter 7. Extraterrestrial Ethics: A Christian Perspective

As we set foot into the cosmic expanse, the ethical implications of space exploration come to the fore. Our understanding of ethics, significantly influenced by religious beliefs and teachings, begins to take on new dimensions. From a Christian perspective, we turn to the teachings of Jesus Christ for guidance in this new frontier.

7.1. A Paradigm Shift in Biblical Interpretations

The Bible has often been interpreted in ways that accord with the worldview and scientific understanding of its interpreters. As our horizons expand to include the cosmos, the interpretation of biblical texts encompassing the universe and mankind must also evolve.

In the book of Genesis, humankind was given dominion over the earth. Traditionally, this has been interpreted as a terrestrial limit placed on humankind's domain. However, as space travel redefines our understanding of 'our domain', it may be necessary to expand our interpretation of this passage. Christian caretaker ethics, influenced by God's call in Genesis to tend the world garden (Genesis 2:15), can now encompass our broader cosmic home.

7.2. Human Life Beyond Earth: Church Viewpoints

Different churches have different perspectives on the idea of human life extending beyond earth. The Roman Catholic Church appears to be the most open, with no official teachings opposing the consideration of extraterrestrial life or human colonization of other planets. Pope Francis stated, "If Martians arrived tomorrow, and they liked us and wanted to be baptised, I would baptise them." This statement reveals an expanded view of the Church's role in the cosmos, reducing the terrestrial restrictions theoretically imposed on it earlier.

Many Protestant churches, especially Evangelical and Pentecostal communities, may have diverse views, given their less centralized nature. Though yet there's an absence of formal discussion on extraterrestrial ethics, it suggests the need for a thorough, respectful, and patient conversation within these communities about the implications of extending human life beyond Earth.

7.3. Space Colonization and the Value of Life

Christian teachings uphold the sanctity and dignity of all life, which drives the conversation on the ethics of tampering with alien ecosystems. Terraforming planets, which may require drastically altering their climates to make them habitable for humans, poses serious ethical questions. Should we, as stewards of God's creation, carry out such drastic alterations, potentially damaging or destroying native alien life, if it exists? The sanctity of life - a Christian ethic - needs to be interpreted from a broader cosmic perspective.

7.4. Distributive Justice in Space

Jesus preached love for all and emphasized treating all individuals as inherently valuable and entitled to basic human rights. He set a precedent for concepts of distributive justice with sayings such as, "To whom much is given, from him, much is expected" (Luke 12:48).

However, the movement into space could lead to new disparities in access and control over resources. As economically advantaged countries or private entities gain power through improved technology and resources, the risk of a cycle of advantage and disadvantage increases. As Christians, we are called to ensure that distribution of resources and opportunities in space adheres to principles of justice, preferential attention to the least advantaged, and equitable access.

7.5. Unity of Mankind in Space

The move from a geocentric to a heliocentric model in the 16th century posed significant challenges to European Christianity's anthropocentric worldview. Now, as we transition from an earth-centred worldview to a cosmos-centred outlook, our conception of mankind's unity must evolve.

In line with Jesus' teachings, which emphasized love for all regardless of race, nationality, or status, the unity of mankind beyond earth needs to be engrained in our space ethics. This unity will not only reach across human differences but also temporal and spatial dimensions, reflecting an infinite love that spans galaxies.

7.6. Creation Care Beyond Earth

The call of Christian stewardship extends to the space environment and our care for any potential life therein. We need to ensure a balance between space exploration and maintaining the space environment so that future generations can also explore, learn, and draw inspiration from it. Debris left behind from space missions, which could interfere with observations or potentially collide with other objects, is a pressing concern that calls for responsible exploration practices.

In conclusion, as we chart our course into the universe, it becomes clear that our ethical boundaries must expand beyond their traditional confines. A robust Christian extraterrestrial ethic will embrace biblical reinterpretation, respect for alien life, distributive justice in space, unity of mankind beyond earth, and creation care beyond earth. As we balance our ambition and responsibility with faith, we can add a harmonious verse to the story of mankind's progress. This expansiveness will not only encompass terrestrial ethics but will embrace a cosmic ethic rooted in divine love, care, and unity, reaching out to the furthest stars and beyond. The frontier may be new, but the principles guiding us remain the same; eternal love, unity, and justice. This balance will ensure that our leap into the cosmos isn't a departure from our ethical foundation but a continuation of our faith journey into new realms.

Chapter 8. Walking in His Steps: Space Travel and Christian Faith

Space. The final frontier. For eons, mankind has cast wondering eyes towards the inky expanse above, yearning for the knowledge of what lies beyond Earth's atmospheric veil. In our relentless pursuit of understanding, we have developed remarkable technology, the culmination of which propels us towards a future where interstellar travel is commonplace; a reality where humans are not confined to Earth, but thrive across galaxies.

8.1. The Bible and Space Exploration

The theology of Christianity offers profound teachings to ponder upon during these trailblazing escapades into the cosmos. Within this intricate blend of cosmic wonder and religious exploration, there lies a central question: How does Christian faith reconcile with the rising era of space travel?

Drawing from the foundations of the Bible itself, humans are cornerstones of God's sublime creation, gifted with intelligence and curiosity. Genesis 1:28 refers to God's command, stating "Be fruitful and increase in number; fill the earth and subdue it. Rule over the fish in the sea and the birds in the sky and over every living creature that moves on the ground." Ever since, mankind has set out to apprehend the world around us. In our time, this mandate has extended to space. We venture into the cosmos, trusting God's Providence, striving to understand more of His creation.

8.2. Jesus Christ: The Ultimate Astronaut

Throughout the New Testament, Jesus Christ is often depicted as a traveler - a celestial voyager who descended from the heavenlies to us. John 1:14 identifies Him as 'The Word' that 'became flesh and dwelt among us.' Christ's mission has always been intrinsically linked to the process of journeying, leading a path into the unknown for humanity, as He did from the heavens to Earth.

Perhaps space explorers are the disciples of the new age, traversing the physical distances as Jesus traversed spiritual chasm - a formerly insurmountable gap now conquered. These astronauts are adhering to the path that Jesus walked: a journey of faith into the unknown, armed with divine hope and love.

8.3. Faith and the Final Frontier

Space travel is indeed a daring leap into uncertainty. Astronauts bound for the distant stars must grapple with the unknown—the knowledge of never returning, the potential danger, and the sheer isolation that the void of space imposes. It is a journey that recalls Jesus's 40 days of wandering in the wilderness or His harrowing, yet redemptive, journey to the cross.

In times of isolation and suffering, Christian astronauts may find comfort and strength from faith. Their trials parallel the trials of Jesus, their courage mirrors His persistence, and their loneliness reflects the solitude He experienced on the cross. By viewing space exploration through the Christian lens, these astronomical voyages are not solely scientific expeditions, but spiritual journeys showcasing faith, hope, and endurance.

8.4. A Creation Consciousness

The dramatic perspective shift experienced by many astronauts, often referred to as the "overview effect," has profound spiritual implications. Viewing Earth from space deepens the realization of humanity's collective existence and interdependence. The beautiful vision of our planet, void of political borders, and teeming with life, imparts a profound oneness with all mankind and creation.

This consciousness mirrors Jesus's teaching of love for all human beings irrespective of their nationality, status, or religion as articulated in the Parable of the Good Samaritan. The "overview effect" gives us a glimpse of God's perspective—where divisions and distinctions fade away under the immense beauty and unity of His creation.

8.5. The Promise of Resurrection in Space Travel

Moreover, space travel encapsulates the Christian concept of resurrection—a rebirth in a strange, new world. As human beings prepare to colonize other planets, we will leave our familiar Earthly lives and start anew in alien terrains, much like the promise of a new and eternal life in Christ after death.

8.6. Leaving Our Comfort Zone

Space exploration is a monumental challenge to our comfort zones. It dares to question our claims on the world, destabilizing our perceived security. In that, it echoes Jesus's teachings: embracing precariousness, renouncing material security, and redirecting our faith unto God's providence.

8.7. The Cosmos as God's Domain

Christianity views the cosmos not as a cold, forbidding void, but as God's domain. "The heavens declare the glory of God," avows Psalm 19:1, asserting that the cosmos is God's canvas, a testament to His power. Space travellers are, then, divine explorers commissioned to explore God's celestial gallery.

Through this odyssey, we are reminded of the infinite love of God, which spans universes, and the promise of unity among all mankind beyond our Earthly bounds. By understanding the intricate connection between Christian faith and space travel, we come to see that these epic journeys into the cosmos are not merely quests for new frontiers, but transcendent pilgrimages wrapped in a journey of faith, destined to bring us closer to our Creator.

Chapter 9. Hope for the Galaxy: Jesus's Teachings in a Space-Age Context

"As we find ourselves on the brink of becoming an interstellar civilization, it is essential to consider the teachings of Jesus Christ and their present relevance in this galactic context. By examining Jesus's lessons within a space-age framework, we may glean new insights into the unity of mankind, infinite love, and the interconnectedness of all things.

9.1. Teachings of Unity

Humankind's journey to the stars is a collective endeavor, reflecting the shared ambition of a race eager to explore its place in the universe. This echoes the teachings of Jesus on the unity of mankind. "As I have loved you, so you too must love one another," he said (John 13:34). Jesus preached of an interconnected brotherhood of humanity, extending a hand of fellowship to all, irrespective of status or reputation.

In a space-age context, as Earth becomes a twinkling speck in the distance, humans must hold fast to the principles of unity and cooperation. Only by truly embracing the vision of Jesus for a unified human family can we hope to prosper in the vast expanse of space.

9.2. Infinite Love Beyond the Stars

The divine love of Jesus is a concept that knows no bounds. It reaches out to the forgotten, empowers the weak, and places a caring hand on the suffering. In his teachings, love is a force as vast as the universe itself, encapsulating everything in its embrace.

As we journey through space, it becomes ever more crucial to hold sacred the concept of infinite love. The isolation and harsh realities of space exploration may challenge our resolve, but the universal love Jesus taught can fill our hearts and habitats, serving as a beacon amidst the alien landscapes.

9.3. The Galactic Parables

Jesus Christ made extensive use of parables to convey spiritual truth. His parables not only talked about the everyday life of people in the 1st century AD but also had a timeless quality, making them relevant even in today's space age context.

Take, for example, the Parable of the Mustard Seed. Jesus used this story to portray the kingdom of heaven as a tiny seed that eventually grows to provide shelter for many creatures. As humanity seeks to establish colonies on other planets, we can draw parallels between our own endeavors and this growing mustard tree. We start small, but with hope and persistence, we can create a new "kingdom" in the extraterrestrial world which can be a haven for many.

9.4. Astro-ethics and Jesus's Teachings

The exploration of new worlds also means the potential discovery of new biospheres and perhaps, sentient life. In such cases, the core teachings of Jesus - respect for all forms of life, kindness, and love - become the foundation for what might be called 'astro-ethics'. If we encounter other forms of life, Jesus's teachings tell us to approach them with love, understanding, and respect.

Jesus believed in the concept of doing unto others as you would have them do unto you (Luke 6:31). If we become the aliens on a distant world, these words may carry a greater depth of meaning than ever before.

9.5. The Unity of Origin and Destination

Jesus's teachings remind us that while we are individuals, we are also intertwined components of a larger whole. Regardless of where we come from or where we go, we are part of the grand narrative of creation and exploration.

As we move beyond Earth and into the broader cosmos, we have the opportunity to embrace this truth on a galactic scale. We might be the citizens of different nations here on Earth, but in the cosmos, we are citizens of the same Universe, bound by the same laws of physics, biology, and, if we choose, the same love that Jesus preached.

9.6. The Hope of Resurrection

The promise of resurrection in Jesus's teachings is a metaphor for hope. It tells us that there is always a chance for rebirth, renewal, and revival, no matter what trials and tribulations we face. This belief can be a powerful motivator for astronauts exploring the galaxy, in the face of adversity and unknown dangers.

The teachings of Jesus Christ map surprisingly well onto the future humankind is reaching for among the stars. His lessons of unity, love, respect, and hope not only help us make sense of our place in the universe but also offer a moral compass guiding our explorations. As we move into the space age, we would do well to keep these teachings close to our hearts, using them to light the path to our cosmic future."

Chapter 10. Spirituality in Zero Gravity: Worship beyond Worldly Bonds

As humanity turns its gaze skyward and takes its first shaky steps into the universe, the shackles of terrestrial confines—gravity, atmosphere, hard soil underfoot—begin to shed. Orbit among the celestial bodies proffers a new perspective, not merely in terms of our physical location, but philosophically as well. We can't help but stand at the mouth of the cosmos and ask ourselves: What does it mean to worship in a realm entirely devoid of earthly constructs?

10.1. The Concept of Infinity and Majesty

There are few things that speak as potently to the scale, dimension, and sheer artistry of God's creation as the tapestry of the cosmos unfurling in all directions. The spectacle of celestial bodies and the divine choreography that they participate in echo the sentiment expressed in Psalm 147:4: "He determines the number of the stars; He gives to all of them their names."

While appreciating the magnificence of the night sky from Earth is a humbling experience in and of itself, immersion in this celestial expanse adds an existential layer to the experience. It's as if space tourists become living verses, mirroring the Apostle Paul's exhortation in Romans 12:1, to "present your bodies as a living sacrifice, holy and acceptable to God."

A central element in Christian worship is the acknowledgement and appreciation of God's majesty. This recognition settles in more profoundly when one floats among the stellar objects of His labor.

Space tourists are poised to perceive the grandeur of Divine design from an unmatched vantage point, enabling them to participate in an uncommon form of worship—a praise that is fundamentally experiential.

10.2. Worship as Unity in Diversity

However, there's more to worship beyond admiration of the universe's grandeur. Recall John 13:35, which states: "By this everyone will know that you are my disciples, if you love one another." This spirit of unity and love personifies the philosophy of Christian worship.

Space travel invites a diversity of cultures, races, and backgrounds into a compact vessel hurtling through the cosmos. This amalgamation can become the embodiment of God's kingdom that cherishes plurality and solidarity. Therefore, as a microcosm of global unity, space tourism takes on a metaphorical mantle of worship by promoting mutual respect, peace, and shared experiences. The ship becomes a cathedral – informal yet sacred, where acceptance is the liturgy and belonging, the hymn.

10.3. Reaching Out to the God of the Universe

Floating in the void invokes a sort of contemplative isolation, erasing the busyness of the world and fostering a transcendent connection. Prayer, a cornerstone of Christian faith, runs parallel to this state of serene isolation. Whether whispered, silent or singsong, thousands of miles above the Earth, each prayer becomes a cosmic dispatch navigating across divine wavelengths.

Such spiritual communion reiterates how God, as the omnipotent Creator, fashions bonds with His creation at any scale or distance.

God's ubiquity becomes more tangible when one prays under cathedral-arch constellation canopies or with Earth's blue visage reflecting in their eyes. Thus, space offers a unique sanctuary whereby humanity can reach out to the God beyond, within, and amidst the cosmos.

10.4. Eucharist Beyond The Earthly Altar

While the celestial perspective may offer novel ways to experience prayer and worship, implementation of specific sacraments like the Eucharist poses several logistical challenges. Yet, Buzz Aldrin shared communion on the lunar surface during the Apollo 11 mission, demonstrating that even this seemingly earthbound ritual can be adapted for a cosmic setting. Precisely measured wine and bread migrated from the earthly altar to the moon's Sea of Tranquillity, introducing a new scale to the Catholic understanding of transubstantiation and Christ's presence.

The Eucharist encapsulates reminiscences of the Last Supper and Christ's sacrifice. Commemorating this in space could symbolize human and divine unity transcending worldly limits. Adjustments must be made to accommodate the physical realities of zero gravity—alterations that can offer a fresh tangibility to this most sacred ritual.

10.5. Towards a Cosmic Theology

As our telescopic reach extends further into the universe, humans yearn to understand how our faith traditions intersect with tomorrow's cosmic reality. As Christians traveling in space explore worship in this extraordinary backdrop, they unlock avenues for understanding God beyond our planetary confines.

Much like theologians of centuries past, who grappled with the reality of an Earth that wasn't the physical center of the universe, we too have before us an opportunity laden with complexity, potential, and utmost fascination. The rise of a "cosmic theology" that integrates scientific understanding and spiritual truths may be awaiting us on the horizon.

In the zero gravity of space, we are challenged to conceive a worship that transcends earthly bonds. How we shape our theology to match our ever-expanding universe will shape the spiritual journey of humanity for centuries to come and possibly set a course for an unparalleled unity beyond the confines of our 'small, blue dot'.

Chapter 11. The Last Chapter: Reimagining Heaven in the Space Age

The concept of heaven, an idyllic paradise beyond the physical realm, has served as the spiritual pinnacle of countless belief systems over the millennia, offering comfort, hope, and a profound sense of connectedness with something greater. Notably, Christian theology perceives heaven as the dwelling place of God, a divine space of eternal life, peace, and joy. Yet, as technology challenges our understanding of the cosmos, it also urges us to expand our conception of what heaven might be like.

11.1. Shades of Heaven in Christian Theology

Christian theologies have traditionally presented varying representations of heaven. Some lean heavily into symbolism and metaphor, while others promote a more literal interpretation. Regardless of their divergences, they consistently depict heaven as a space removed from earthly constraints, where good triumphs and God's love is unceasing.

As we traverse the threshold of the Space Age, our perspectives of the universe are being transformed and broadened. Earlier, the sky served as a comprehensible and comforting canvas on which we drew our beliefs. Now, as we receive intriguing glimpses of the vast expanse beyond, we're compelled to reexamine and reimagine these paradisiacal representations of heaven.

11.2. Heaven and Human Cognition

It might be tempting to dismiss these conceptual shifts as merely scholarly explorations, but consider this: our conceptions of heaven are woven intrically into our perceptions of morality, purpose, and destiny. It shapes how we engage with ourselves, our neighbors, and our environment. Hence, any shift in such a fundamental conception will naturally influence humanity's spiritual direction.

For many, heaven is the opposite of the hardship, suffering, and impermanence we witness in the world. It is a manifestation of our deepest longings – for unconditional love, harmonious existence, and eternal life. The symbolism of heaven also empowers us to bear the trials of life with grace and courage, encouraging us to echo divine goodness in our earthly interactions.

11.3. Heaven: A Cosmic Realm

As we gain greater awareness of the universe's vastness, our depictions of heaven are increasingly being seen through a cosmic lens. Could heaven be a tangible place within this universe? Could it exist in another dimension or reality that our current scientific understanding hasn't yet grasped?

Exploring such questions might seem daunting or even sacrilegious to some. But consider this: perhaps, our space explorations are not moving us further from God or heaven but, instead, closer to them. Such a perspective doesn't seek to refute or replace traditional beliefs but only to extend them, thereby aligning with our unfolding celestial journeys.

11.4. Jesus and the Space Age

In this era of technological advancements and space tourism, we might find fresh insights within old scriptures. Jesus Christ, the heart of Christian theology, brought a radical message of universal love and unity. He preached that the Kingdom of Heaven is not just a place far removed, but it is also here within us and around us.

This aligns seamlessly with our voyage into the cosmos – as we step out into space and reach out to celestial bodies, are we not scaling the walls of perceived separateness? Are we not moving towards a profound realization of interconnectedness, akin to the unity Jesus espoused?

11.5. Unity Beyond Earth

As we venture further into space, we find unity - not just with fellow spacefarers, but with all of humanity, from whom we can meaningfully perceive our interconnectedness. The 'overview effect,' experienced by many astronauts describes this phenomenon.

From a vantage point outside of the Earth's surface, petty disagreements and human-made divisions seem insignificant. We start to recognize the intricate web of life and the force of love that underpins all existence, resonating closely with Jesus's teachings.

11.6. The Promise of Eternal Life

We understand that the journey to space is still fresh and fraught with many challenges, including ethical questions about our right to venture and settle on heavenly bodies. However, with careful deliberation and humble learning, we may indeed find in this journey signs of the infinite God of love that Jesus demonstrated.

Eventually, this quest can also open up a discourse about eternal life. The concept of immortality or life extension is rapidly moving from the realm of speculative fiction to scientific possibility. However, science and theology can come together to recognize that eternal life is less about physical continuity and more about demonstrating undying commitment to compassionate stewardship of all beings.

In conclusion, as we venture farther into the cosmos, our notions of heaven need not necessarily become more obscure or fragmented. On the contrary, this journey might illuminate our hearts with deeper understanding and conviction as we draw inspiration from the teachings of Jesus and the wonders of space. Our voyage to the stars might prove how truly close we are to the heaven we seek. Ultimately, the lines between space exploration, theology and human ambition may not be as clear as we imagined. The interplay between them allows for a continuous dialog, and as we learn more about our universe, we are allowed the opportunity to further understand our shared values and our common destiny. Heaven, in such an integration of perspectives, is not just a glorious end but also a beautiful beginning, a constant guide shining in the vast expanse of our explorations.